Bibliographic information published by the German National Library:

The German National Library lists this publication in the National Bibliography;
detailed bibliographic data are available on the Internet at http://dnb.dnb.de .

Imprint:

Copyright © 2019 GRIN Verlag
Print and binding: Books on Demand GmbH, Norderstedt Germany
ISBN: 9783668975750

This book at GRIN:

https://www.grin.com/document/489009

Saharsh Khicha

Inter-specific and Intra-specific Competition in Plants

GRIN Verlag

SAHARSH KHICHA

THE UNIVERSITY OF HONG KONG

Interspecific and Intraspecific Competition in plants.

Research Question: Does decreasing distance (70mm, 50mm, 30mm, 10mm) of the wheat (*Triticum aestivum*) and chickpea (*Cicer arietinum*) seeds sown in the presence (5% solution of urea) and absence of urea effects the shoot length (mm) due to inhibitory effects of intra specific and inter specific competition between them?

Word Count: 3952

Table of Contents

1. Introduction

Crop production in India is labour intensive with limitation of scientific techniques. Unawareness in farmers for correct distance of sowing seed leads to low production of crops. This is due to overlapping of niche and thus the realised niche of a species which is the actual mode of existence resulting from adaptation and competition is developed[1]. A research was undertaken by **Jianbo Shen, Lixing Yuan et al.** in Journal of Experimental Botany which acknowledged various factors that affect the crop production and solutions to overcome competition[2]. The research would make farmers aware of a major factor like competition, that should be considered while crop production.

A crop is produced when a seed is germinated and plumule and radicle grow. Among various factors that affect the shoot length of a plant are light, oxygen, soil fertility, moisture supply, nutrients supplied etc. competition is least considered. Length of the shoot system depends upon how better the plants can overcome the competition to obtain above resources. If plants are of different species they have different resource requirement but if plants are of same species their resource requirement is very much similar and hence lead to tolerance and stress[3]. Urea was chosen as a nutrient source as according to an exploration by **Rein Aerts**, the competition is also dependent on the nutrient provided to the plant. An investigation on effect of distance between plants on the development of cassava plant tuber was conducted by **Nereu Augusto Streck et al.** which resulted in high yield per plant with lower densities[4]. So this investigation was planned to work on the different types of competition and solve the problem of the farmers suffering from shortages of crop production. Hence, the research question was framed:

"Does decreasing distance (70mm, 50mm, 30mm, 10mm) of the wheat (*Triticum aestivum*) and chickpea (*Cicer arietinum*) seeds sown in the presence (5% solution of urea) and absence of urea effects the shoot length (mm) due to inhibitory effects of intraspecific and interspecific competition between them?"

1.1 Competition

It is the struggle between the plants or groups of plants for nutrients, food and pollinators in order to grow and survive[5]. Competition can occur between the plants of same species called intraspecific competition and it

---

[1] Minka Peeters. IBID Biology. 4th ed. Melton, Australia: pg 448 IBID Press, 2008. Print.

[2] Fan, M. et al. "Improving Crop Productivity And Resource Use Efficiency To Ensure Food Security And Environmental Quality In China." Journal Of Experimental Botany, vol 63, no. 1, 2011, pp. 13-24. Oxford University Press (OUP), doi:10.1093/jxb/err248. 3 Minka Peeters. IBID Biology. 4th ed. Melton, Australia: pg 454 IBID Press, 2008. Print.

[4] Streck, Nereu Augusto et al. "Efeito Do Espaçamento De Plantio No Crescimento, Desenvolvimento E Produtividade Da Mandioca Em Ambiente Subtropical." Bragantia, vol 73, no. 4, 2014, pp. 407-415. Fapunifesp (Scielo), doi:10.1590/1678-4499.0159.

[5] Craine, Joseph M., and Ray Dybzinski. "Mechanisms Of Plant Competition For Nutrients, Water And Light." *Functional Ecology*, vol 27, no. 4, 2013, pp. 833-840. *Wiley-Blackwell*, doi:10.1111/1365-2435.12081.

also occur between plants of different species known as the interspecific competition[6]. Intraspecific competition is usually more intense than the interspecific competition[7]. Generally to decrease the intraspecific competition between the plants, they have a technique of seed dispersal[8].

1.1.1 Interspecific competition

It is the competition which occurs between two different plants of dissimilar genetic constituents as if the seeds of different plants are sown closer to each other then they will affect growth of each other[9]. The interspecific competition may decrease or increase the growth rate of a particular plant. It is dependent on the availability of the nutrients when sown in natural conditions.

1.1.2 Intraspecific competition

When a member of same species compete for the food, shelter, water and mates then this competition is known as intraspecific competition[10]. In case of plants when two or more seeds of same species interfere with each other and hence affect the growth rate of the plant it is known as intraspecific competition in plants.

2. Background Information

In this investigation both inter and intraspecific competition are developed through decreasing distances (mm) between sown seed and exposure to urea for contain group of both plant seeds after germination.

2.1 Wheat (*Triticum aestivum*)

Wheat is one of the most abundant crop grown and consumed in India. So it was decided to take wheat to provide the data of a crop which is more useful to the farmers. Production of wheat in India as per the year 2017, is 96000 (1000 MT)[11]. So this seed was chosen for the investigation as it is easier to grow and do not need enough maintenance. Life cycle of the wheat seeds is for the total of approximate 120 days. 4 -5 days

[6] Craine, Joseph M., and Ray Dybzinski. "Mechanisms Of Plant Competition For Nutrients, Water And Light." *Functional Ecology,* vol 27, no. 4, 2013, pp. 833-840. *Wiley-Blackwell,* doi:10.1111/1365-2435.12081.

[7] Beals, M., et al. "INTERSPECIFIC COMPETITION: LOTKA-VOLTERRA." INTERSPECIFIC COMPETITION, 12 Dec. 2017, www.tiem.utk.edu/~gross/bioed/bealsmodules/competition.html.

[8] Minka Peeters. IBID Biology. 4th ed. Melton, Australia: pg 447 IBID Press, 2008. Print.

[9] Skálová, Hana et al. "Effect Of Intra- And Interspecific Competition On The Performance Of Native And Invasive Species Of Impatiens Under Varying Levels Of Shade And Moisture." *Plos ONE,* vol 8, no. 5, 2013, p. e62842. *Public Library Of Science (Plos),* doi:10.1371/journal.pone.0062842.

[10] *Fcnym.Unlp.Edu.Ar,* 2018, http://www.fcnym.unlp.edu.ar/catedras/ecopoblaciones/TP/Begon%20Capitulo%205.pdf.

[11] "India Wheat Production by Year." India Wheat Production by Year (1000 MT), www.indexmundi.com/Agriculture/?country=in&commodity=wheat&graph=production.

are required for the wheat seeds to germinate. This is the time when the seeds are fully developed into a wheat plant[12]. A young seedling is formed 4-5 days after the germination. This is a two leaf stage[13].

2.2 Chickpea (*Cicer arietinum*)

Chickpea seeds are one of the most important Rabi crop grown in India. India is one of the major country for the growth of chickpea seeds. India produces 8 million tons of chickpea seeds which is 70% of the world's total production[14]. Chickpea seeds were chosen as its shoot grows rapidly. Complete Life Cycle of chickpea seeds takes place in approximately 100 days. The duration of rapid development of the chickpea seeds is before the flowering stage of the seeds[15].

2.3 Urea

Urea is used as a fertilizer to grow various crops. Urea is formed by two amide groups ($-NH_2$) which is bonded to a central carboxylic carbon ($-C=O$)[16]. Urea acts as an additional nutrient in the germination of a seed. It helps in providing nitrogen to the soil. Nitrogen is the important component which affects the growth of the plant[17]. It increases the formation of chlorophyll which helps in photosynthesis due to which the plant's growth is triggered[18]. Nitrogen is the main fertilizer for the plants to provide nutrient[19]. So it was decided to use urea as a source for nutrient to the plant.

20

2.4 Medium to grow seed

Sand was used to germinate and grow the seeds properly. As different quality of soil may vary the results in the shoot length of the seeds[21]. As soil contains some extra nutrients which will affect the shoot length of the

[12] "How to Grow Wheat | Guide to Growing Wheat." Organic & Heirloom Seeds from Heirloom Organics, www.heirloom-organics.com/guide/va/guidetogrowingwheat.html.

[13] Growth and Development Guide for Spring Wheat, www.extension.umn.edu/agriculture/small-grains/growth-and-development/spring-wheat/.

[14] "ALL INDIA COORDINATED RESEARCH PROJECT ON CHICKPEA." ALL INDIA COORDINATED RESEARCH PROJECT ON CHICKPEA, www.aicrpchickpea.res.in/.

[15] Nogueira, S.S.S., et al. Growth Analysis of Chickpea (Cicer Arietinum L.). Dec. 1994, www.scielo.br/scielo.php?script=sci_arttext&pid=S0103-90161994000300008.

[16] "Urea Formula." Math, www.softschools.com/formulas/chemistry/urea_formula/420/.

[17] "Fertilizer Urea." Fertilizer Urea : Nitrogen : Nutrient Management : Agriculture : University of Minnesota Extension, www.extension.umn.edu/agriculture/nutrient-management/nitrogen/fertilizer-urea/.

[18] "Nitrogen." Efficient Fertilizer Use Guide Nitrogen | Mosaic Crop Nutrition, www.cropnutrition.com/efu-nitrogen.

[19] Acevedo, E., et al. Wheat Growth and Physiology. 2002, www.fao.org/docrep/006/y4011e/y4011e06.htm.

[20] "Physical Properties Of Urea." *Sciencing*, 2018, https://sciencing.com/physical-properties-urea-6369247.html.

[21] "Factors Affecting Germination of Seeds." Factors Affecting Germination of Seeds | TutorVista.com, www.tutorvista.com/biology/factors-affecting-germination-of-seeds.

plant. So sand was chosen. Sand must be made free from any impurities such as stones, leaves and other impurities.

2.5 Seed Germination

Seed Germination is the growth of seeds into young plant or seedlings in the presence of sunlight and water[22]. It is the period of the growth of embryo. It is the stage where the seed begins to grow in a plant. In this phase a primary root comes out of the seed known as the radicle of the plant. The plumule of the seeds also emerges at this time and it is developed into the shoot system of the plant[23].

3. Hypothesis

3.1 Null hypothesis

With the decrease in the distance between the seeds there would be no effect on the plumule lengths of the seeds for both the interspecific and the intraspecific competition whether the experiment is performed in presence or absence of urea. So when the t- test will be conducted the probability to accept the null hypothesis would be less than 5% or 0.05 ($p < 0.05$)

3.2 Alternate hypothesis

With the decrease in the distance between the seeds there would be decrease in the shoot lengths of the seeds for both the interspecific and the intraspecific competition. And the shoot length will be greater in the presence of urea solution as extra nutrients are provided to the plant seeds. Moreover, a plant will perform well under interspecific competition rather than intraspecific competition[24]. Therefore, when the t- test would be conducted the probability that the alternate hypothesis would be accepted is greater than 95% or 0.95. ($p > 0.95$)

4. Variables

4.1 Independent variables

1. Availability of urea – the concentrations of urea was changed by me to see the effect of it on the plumule length of the seed after germination. Its value is 0% and 5%
2. The distance between the seeds – distance (mm) between the seeds was decreased and observed in order to see the interspecific and intraspecific competitions.

---

[22] Chandrasekara, Dinara. "Germination | Seed Germination for Kids | Science Lessons for Kids." k8schoollessons.Com, 29 Aug. 2017, k8schoollessons.com/germination/.

[23] Burchill, Shirley. "REPRODUCTION IN PLANTS." The Open Door Web Site : Biology : Reproduction in Plants : Conditions Needed for the Seed to Germinate, www.saburchill.com/chapters/chap0048.html.

[24] Skálová, Hana, et al. "Effect of Intra- and Interspecific Competition on the Performance of Native and Invasive Species of Impatiens under Varying Levels of Shade and Moisture."PLoS ONE, Public Library of Science, 2013, www.ncbi.nlm.nih.gov/pmc/articles/PMC3651240/.

4.2 Dependent variables

1. Shoot length of the seeds – the shoot length of seed were measured which was varied with the absence and availability of urea and the decreasing distance between the seeds.

5. Methodology

5.1 Planning to introduce competition between growing plant

The distance in the seeds was decreased to 70mm, 50mm, 30mm, and 10mm for imposing competition. Also a source of nutrition must be provided to the seeds to grow. Urea was used as a source of nutrient for the seeds to grow as nitrogen acts as a main fertilizer to the plant.[25] It was decided to perform the setups in the presence and absence of urea for all the distances and then measure the trend in the change in the shoot length of the seedlings after 12 days of the investigation.

5.2 Seed Selection

The seeds of chickpea and wheat are first soaked in water for a night so that it would overcome the initial dehydration of the seeds so that there would be no deviation in the results. Next day, observe if the seed sinks then they are viable and if it floats then that seeds are not fresh and healthy enough[26]. After soaking them in the water the seeds are now ready to grow.

5.3 Seed Bed (sand bed) preparation

In a bucket sand was filled and water was poured in the bucket. The mixture was stirred and then it was filtered to separate out stones and other impurities from the sand so that it would be fresh to use for the investigation. After filtering the sand, the sand was kept in the sunlight for 5-6 hours to let it dry and the sand is ready to use. 1-2 drops of vinegar were used to sterilise each disposable bowl of 100 mm diameter and this was done for all the 80 seed bed container[27].

5.4 Seed laying for the interspecific and intraspecific competition

The temperature of seed bed was maintained around 18- 24 ∘C as the setup was placed in an air conditioned room temperature as it affects the rate and the percentage of the seeds that grow and would affect the shoot length of the plant[28]. The soil moisture was also kept constant throughout the investigation by supplying 40 ml of water to each setting[29]. As due to water stress and production of gibberellic acid in the seed there would

[25] "Nitrogen." Efficient Fertilizer Use Guide Nitrogen | Mosaic Crop Nutrition, www.cropnutrition.com/efu-nitrogen.

[26] "How to Know If Garden Seed Is Viable." Horticulture, 29 Jan. 2013, www.hortmag.com/weekly-tips/propagation/how-to-know-if-garden-seed-is-viable.

[27] Tooley, Jill. "How To Clean Your Reusable Water Bottle Or Travel Mug". *Promotional Products Blog | Quality Logo Products (QLP)*. Web. 2 Dec. 2016.

[28] "Gilbert, Scott. "Germination". *Ncbi.nlm.nih.gov*. Web. 2 Dec. 2016.

[29] Tilley, Nikki. "What Is Soil Temperature: Learn About The Ideal Soil Temperatures For Planting." Gardening Know How, 26 Nov. 2016, www.gardeningknowhow.com/garden-how-to/soil-fertilizers/determining-soil-temperature.htm.

be affect in the nutritional status of seed[30]. The seed bed was prepared from the filtered sand by measuring 200gm equally in all the sterile 100 mm diameter seed bed container for the investigation. The seeds which were soaked in water were now collected for the investigation. 80 settings were used for the investigation and labelled according to their distances. For the intraspecific competition of chickpea seeds, the seeds were kept at the appropriate distances i.e. 70mm, 50mm, 30mm, 10mm measured with the help of measuring scale. Similarly, it was done for the setting of intraspecific competition of wheat seeds and for the interspecific competition between the chickpea and wheat seeds. The setups were then covered with the plastic wraps to prevent any loss of moisture from the soil and the amount of light is also controlled as light of different wavelengths can change the rate of growth and germination[31]. So, all the plants were kept away from the sunlight and were kept under the room light so that equal intensity and wavelength of light is available to all the setups. Distances for the investigation was selected with a difference of 20 mm to see the significant difference on plant's germination and growth due to distance and availability of urea. The setups were left for 5 days to germinate and then the solution of urea was added to those which were labelled as in the presence of urea. Then further the setups were left to grow for 6 days. The time allotted to the plants to germinate and grow should be kept constant so as to reduce the percentage error during the investigation.

5.5 Preparation of urea solution and its use

Among the 80 setups 50% were planned to be grown in presence of urea for both wheat and chickpea plants so 5% urea solution was prepared. 5% concentration of urea was chosen as per the preliminary lab by dissolving 5mg of urea in 100 ml of water. Now to make the solution of 600 ml, 500 ml of water was added to the solution of urea. The urea solution was added after the germination of the seeds and then for 6 days it was allowed to grow and then the effects was observed. Make sure that the time allotted for the germination of all the seeds is the same so that appropriate results could be observed (5 days). Similarly, the time allotted to the growth of the shoot length is also fixed i.e. 6 days.

5.6 Measuring the shoot length

After that the shoot lengths of each seed sown was measured by a ruler of 30 cm. The same 30 cm ruler was used throughout the investigation to measure the varied shoot length of the seeds so that there would not be any error in measuring the length of the shoot. The above process conducted for chickpea will also be done for observing the intra specific competition between wheat and for the interspecific competition between wheat and chickpea.

---

[30] Kaya, C., Levent Tuna, A. & Alfredo, A.C.A. Acta Physiol Plant (2006) 28: 331. https://doi.org/10.1007/s11738-006-0029-7

[31] "The Effect of Light on Germination and Seedlings." Thompson & Morgan, www.thompson-morgan.com/effect-of-light.

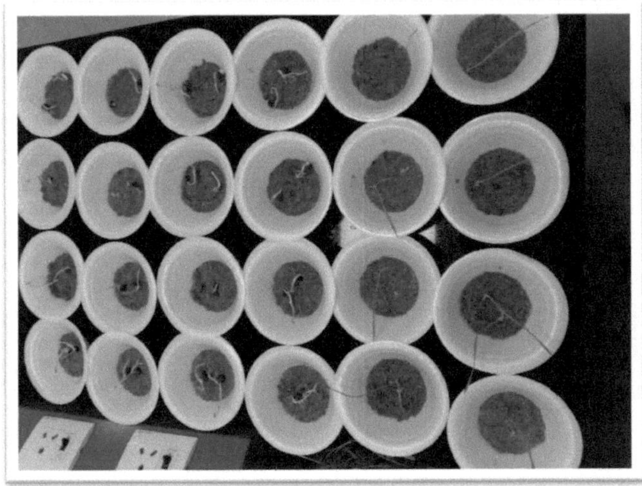

Figure 1: 5 trials for the intraspecific competition for the wheat and chickpea seeds at different distances.

6. Data Collection

As the experiment was done keeping all the variables in mind the shoot lengths of the seeds were collected as shown in figure 1. The figure shows the intraspecific competition between the seeds for different distances. After 5 days of germinating all the seeds under same condition and then adding urea to the respective seeds and letting it grow for 6 days the data was collected. Using a measuring scale of 30 cm the shoot length of the seeds was measured and recorded. The shoot length of the seeds was measured in millimetres. The data was then processed in the table according to the type of competition and availability of urea. Furthermore, the data was then processed and evaluated by conducting t-test and calculating average shoot length. Conducting the preliminary lab, it was observed that the shoot lengths do not have large variations for same distance and concentration. So, a minimum of five trials for each value were taken. Different data tables were made for different type of competition and seeds incorporating all the trials. Another table was developed in which the calculated value of the average of all the trials of the value of the shoot length of the seed differently for interspecific and intraspecific competition. The error bars were also calculated and incorporated in the graphs. Then, graph of the average shoot lengths of the seeds in the absence and the presence of the urea in which they were grown was plotted to analyse the trend of the competitions. The linear regression equations obtained from the graph were then tabulated and analysed. A T-test was also done to check which hypothesis would be accepted.

7. Data Table

Table 1: Average shoot length of the wheat (**_Triticum aestivum_**) and chickpea (**_Cicer arietinum_**) plants for the interspecific competition shown with the decreasing distance (mm) and in the presence (5%) and absence of Urea.

seed type	distance (mm)	Urea (5% conc.)	shoot length of chickpea and wheat seeds ($\pm$ 0.5 mm)				
			1	2	3	4	5
chickpea seeds	70	Presence	105.50	106.00	104.50	105.50	106.00
		Absence	103.50	104.00	104.50	103.50	104.00
wheat seeds		Presence	184.00	184.50	185.50	185.50	184.50
		Absence	181.50	181.00	180.50	182.00	182.50
chickpea seeds	50	Presence	99.00	99.50	99.00	98.50	98.50
		Absence	97.50	98.00	97.00	97.50	97.50
wheat seeds		Presence	171.50	171.00	170.50	170.50	171.50
		Absence	168.50	168.00	167.50	168.50	169.00
chickpea seeds	30	Presence	93.00	94.00	94.50	93.50	93.50
		Absence	92.50	92.00	91.50	92.00	92.00
wheat seeds		Presence	162.00	162.50	163.00	162.50	163.00
		Absence	160.00	159.50	159.50	159.00	159.00
chickpea seeds	10	Presence	87.50	87.00	88.00	87.00	87.50
		Absence	84.00	83.00	84.00	83.50	83.50
wheat seeds		Presence	155.50	154.50	155.00	156.00	154.50
		Absence	152.00	151.50	151.00	152.00	151.50

Table 2: Average shoot length of the chickpea (***Cicer arietinum***) plants for the intraspecific competition shown with the decreasing distance (mm) and in the presence (5%) and absence of Urea.

distance (mm)	urea (5% conc.)	shoot lengths of chickpea seeds (± 0.5 mm)									
		seed 1					seed 2				
		1	2	3	4	5	1	2	3	4	5
70	presence	103.50	102.50	102.50	101.00	101.50	102.00	102.5	102.00	102	102.50
	absence	101.00	100.50	100.50	101.50	99.50	103.50	103	102.00	102.50	103.00
50	presence	97.00	96.50	97.00	96.50	96.50	97.50	98.00	97.00	97.50	98.00
	absence	95.50	95.50	95.00	95.00	96.00	96.50	96.50	96.00	97.00	97.50
30	presence	90.50	91.00	90.00	91.00	90.50	89.50	90.00	90.50	89.50	90.50
	absence	89.50	88.50	89.50	89.00	87.50	89.50	88.50	88.00	89.00	88.50
10	presence	80.50	79.50	80.50	81.00	81.50	82.00	81.50	80.50	81.00	81.50
	absence	79.00	78.50	79.50	79.50	79.00	79.50	80.00	80.00	79.00	79.50

Table 3: Average shoot length of the wheat (***Triticum aestivum***) plants for the intraspecific competition shown with the decreasing distance (mm) and in the presence (5%) and absence of Urea.

distance (mm)	urea (5% conc.)	Shoot lengths of wheat seeds (mm)									
		seed 1					seed 2				
		1	2	3	4	5	1	2	3	4	5
70	presence	180.00	180.50	181.00	180.50	181.00	180.50	181.00	181.50	181.50	180.50
	absence	179.50	179.50	178.50	179.00	178.50	178.00	178	177.50	177.00	177.50
50	presence	167.50	167.00	166.50	167.00	167.50	166.50	166.00	166.50	167.00	167.00
	absence	165.50	165.00	165.00	164.50	165.50	165.00	166.00	165.50	166.50	166.00
30	presence	154.50	154.00	154.50	155.00	155.50	153	154.00	153.50	153.50	154.00
	absence	151.50	150.00	150.50	151.00	150.50	150.00	150.50	149.50	150.00	149.50
10	presence	143.50	140.50	142.00	145.50	144.00	144.00	143.50	144.00	143.00	144.50
	absence	141.00	140.00	139.5	141.00	140.50	142.00	142.50	141.00	141.50	141.00

Table 4: Average values for the Shoot lengths of the wheat (***Triticum aestivum***) and chickpea (***Cicer arietinum***) seeds grown in decreasing distance (mm) and also in the availability of the urea and categorised into interspecific and intraspecific competition between them.

distance (mm)	Urea (5% conc.)	Average Shoot lengths of chickpea seeds (mm)		Average shoot lengths of wheat seeds (mm)	
		Interspecific	Intraspecific	Interspecific	Intraspecific
70	presence	105.5	102.2	184.8	180.8
70	absence	103.9	101.7	181.5	178.3
50	presence	98.9	97.15	171	166.9
50	absence	97.5	96.05	168.3	165.5
30	presence	93.7	90.3	162.6	154.2
30	absence	92	88.75	159.4	150.3
10	presence	87.4	80.95	155.1	143.5
10	absence	83.6	79.35	151.6	141

8. Graphs

Graph 1: Average Value of the shoot lengths for the interspecific and intraspecific competition of the wheat and chickpea seeds grown in the soil at a distance of 70 mm, 50 mm, 30 mm and 10 mm in the absence (0% concentration) of urea.

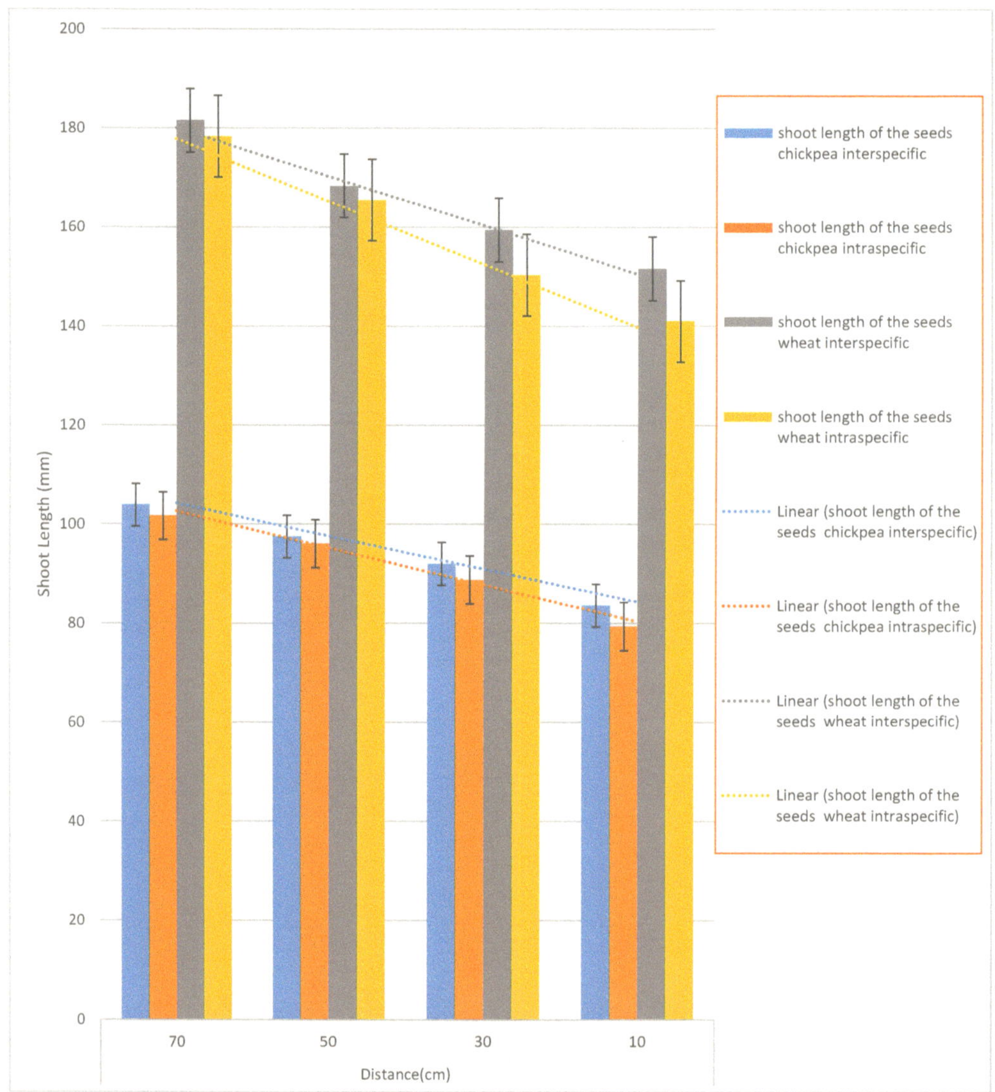

Graph 2: Average Value of the shoot lengths for the interspecific and intraspecific competition of the wheat and chickpea seeds grown in the soil at a distance of 70 mm, 50 mm, 30 mm and 10 mm in the presence (5% concentration) of urea.

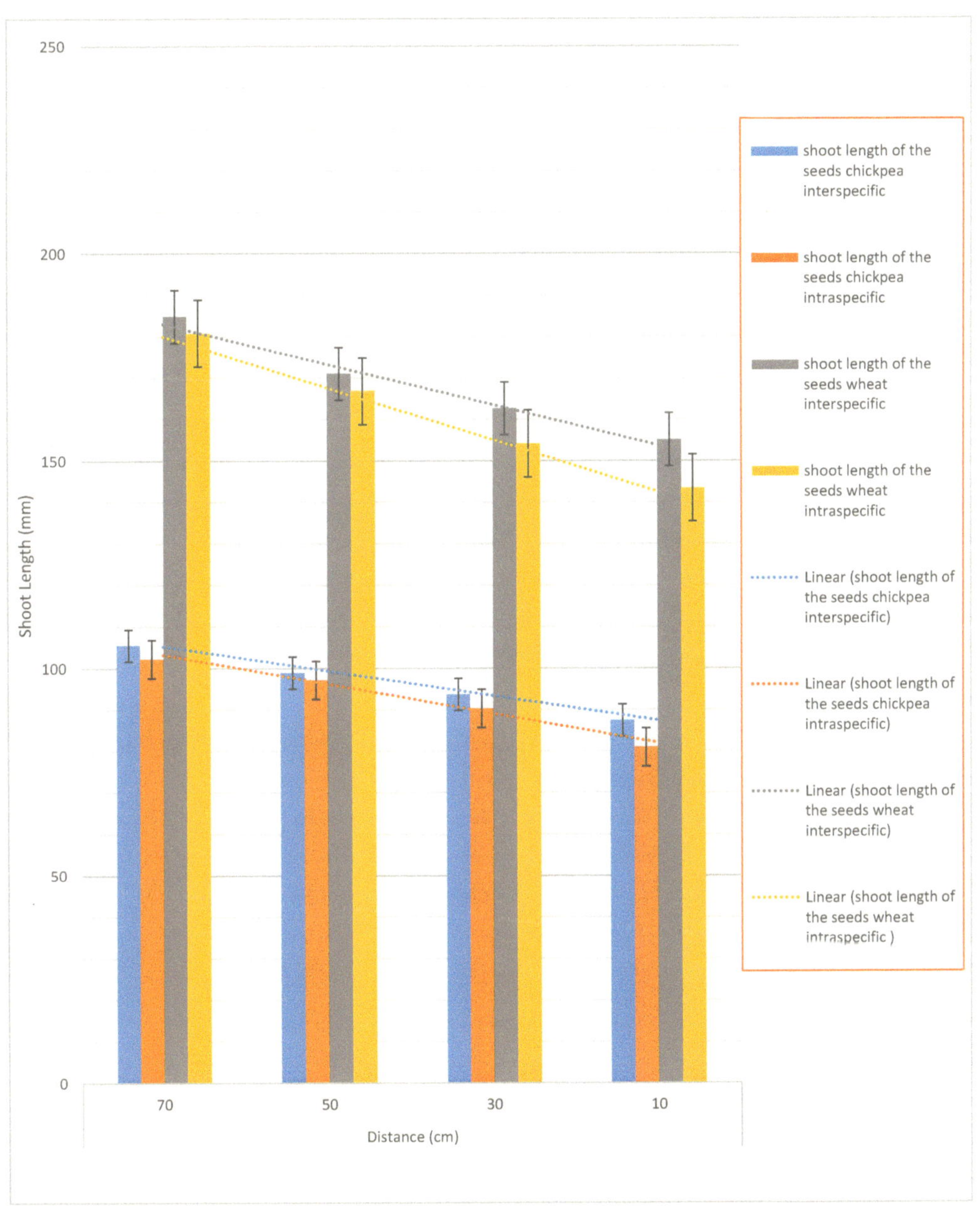

Table 5: The table of the equation of the linear best fit of the graph for the seeds in the presence and absence of urea

Urea	Equations of the Trend lines			
	corn		wheat	
	interspecific	intraspecific	interspecific	intraspecific
Presence (5%)	-5.95x + 111.25	-7.058x + 110.29	-9.75x + 192.75	-12.475x + 192.5
Absence	-6.64x + 110.85	-7.435x + 110.05	-9.86x + 189.85	-12.705x + 190.53

9. Results

With the processing of data, it was observed that the shoot length of the plant was greater even in 10 mm in the presence of the urea i.e. 5% concentration of urea. As urea was added to chickpea and wheat seeds kept at 10 mm from each other the value of the shoot length kept on increasing in both the seeds (chickpea as well as wheat) (Table 5). The shoot length in the absence of urea for the intraspecific competition in the chickpea seeds increased by 1.6 mm in the presence of urea. The shoot length in the absence of urea for the intraspecific competition in the wheat seeds also increased by 2.45 mm in the presence of urea.

Moreover, it has been seen that intraspecific competition is more effective than interspecific competition in both cases i.e. in chickpea seeds as well as wheat seeds (Graph 1 and 2 and table 5). It was observed that as the competition between the plants increases the shoot length of the plants is affected due to restriction of growth. The average shoot length of chickpea seeds in the absence of urea for 10 mm was 9.4 lesser than that when the distance is increased to 30 mm (Table 4). In case of interspecific competition of chickpea seeds as the distance increases from 10 mm to 30 mm the shoot length increases by 8.4 mm (Table 1). Moreover, in the case of interspecific competition of the wheat seeds if the distance is increased from 10 mm to 30 mm then the shoot length is moving just by 7.8 mm (Table 1).

The shoot length of wheat seeds was more than the shoot length of chickpea seeds might be due to excessive food stored in it. Table 5 shows the trends of the slopes of the graphs 1 and 2. The data supports that intraspecific competition was more effective than interspecific competition. The slope for the best fits for intraspecific is more negative than the interspecific competition in the absence and presence of urea. More negative means that the shoot length is decreasing by a greater ratio. As the slope for the intraspecific competition in the absence of urea was 0.759 more negative than the interspecific competition. Therefore, sowing same seeds at same place is more dangerous than different seeds.

10. Discussion and Evaluation

According to the competitive exclusion principle, no two species competing for the same resources in the same environment cannot stably exist[32]. The present investigation surrounds the same principle. The investigation dealt with the effect of availability of urea and the effect of decreasing distances between the seeds on the shoot length of wheat (*Triticum aestivum*) and chickpea seeds (*Cicer arietinum*) for 12 days. The results from this investigation was observed that intraspecific competition has greater effect on the shoot length of the plant than the interspecific competition. The presence of urea (5% in concentration) also catalysed the growth of the shoot length of the plant as it was observed that the average shoot length of both the seeds in both the competition was greater in the presence of urea than in the absence of urea. Therefore, the shoot length of the wheat and chickpea plants increased in both interspecific and intraspecific competition as it was sown in the presence of urea.

In order to find the competition between the plants the seeds were sowed at different distances and also this investigation was able to check the use of fertiliser for the growth of the crop. The investigation was also able to prove that in the presence of urea there was rapid growth in the plant as external nutrients were provided. A similar investigation was conducted by the scientist Peter Stoll, Deborah R. Vogt and David J. Murrell in which **they studied the competitive response of the plants in the interspecific and the intraspecific competition and its effect as the distance increases**[33]. This experiment proposed that the competitive response decreased as the distances were increasing. So the results were supported by this investigation. Thus, the results were totally supportive with the alternate hypothesis provided at start of the investigation.

The average shoot length of the wheat seeds was 68.9 mm greater than the shoot length of chickpea seeds for 30 mm distance between the seeds in the absence of urea for the interspecific competition. Researching, it was found that this is because of the biomass of the seeds[34]. As the biomass in the seeds is different the growth rate is different due to which the shoot lengths vary between the wheat and chickpea seeds. Therefore, wheat

---

[32] "Principle of Competitive Exclusion." Encyclopaedia Britannica, www.britannica.com/science/principle-of-competitive-exclusion.

[33] Vogt, Deborah & Murrell, David & Stoll, Peter. (2009). Testing Spatial Theories of Plant Coexistence: No Consistent Differences in Intra-and Interspecific Interaction Distances. The American Naturalist. 175. 73-84.

[34] Houghton, Jennie, et al. Proceedings of the Royal Society B: Biological Sciences, The Royal Society, 7 July 2013, www.ncbi.nlm.nih.gov/pmc/articles/PMC3673064/.

seeds had greater height of the plant when kept within the same time. These observations do not affect the trend of the competitions in both the seeds. The trend for the specific competition is similar for both the seeds.

Furthermore, t- test was done to show the differences between the shoot lengths of the plants in the presence (5% concentration of Urea) and absence of urea (table 2-3). Although the test conducted for the experiment didn't yield a positive result for any significant statistical difference as the value calculated was very small than the degree of freedom, $p < 0.05$ (P- value shown in the table in appendix-iii). As the calculated value for the t- test for 10 mm of the distance between the seeds is 3.04×10^{-5} and the P- value is 2.101 which is very large then the calculated value. So, the t- test says that there is 95 % probability that the results obtained for the effect of concentration of urea on the shoot lengths will support the null hypothesis. Furthermore, there is only 5% chances that the results will support alternate hypothesis. So, the t- test done did not yield the results supported by this investigation. One of the major problems for the failure of the t- test may be due to the insufficiency of data statistically.

Overall this study was successful in proving the alternate hypothesis correct. Through conducting the experiment and analysing of the data to form a trend has led to focus the problem of the farmers. The investigation proved that as the distance between the seeds decreased the shoot lengths were comparatively lesser and in the presence of urea the shoot lengths were comparatively greater. Hence this investigation provided with the appropriate distances between the seeds and number of fertilisers that should be added to the crops in order to get maximum yield. The investigation was successful in making the farmers aware about the major factors which affect the crop yield.

11.   Conclusion

An exertion was made to discover the inhibitory impact made due to increase in the interspecific and intraspecific competition on the development of the shoot length of wheat and chickpea seeds. Hence, a research question was encircled **"Does decreasing distance (70mm, 50mm, 30mm, 10mm) of the wheat (*Triticum aestivum*) and chickpea (*Cicer arietinum*) seeds sown in the presence (5% solution of urea) and absence of urea effects the shoot length (mm) due to inhibitory effects of intra specific and inter specific competition between them"?** As it is discussed above that the effect of intraspecific competition is greater than that of the interspecific competition as it is shown through the data that the value in the interspecific competition of the shoot length id shorter than the value of the shoot length in the intraspecific competition.

Data obtained from the investigation exhibit inhibitory effect of the presence and absence of the urea on the shoot length of the wheat (*Triticum aestivum*) and chickpea (*Cicer arietinum*) seeds. Urea acts as a catalyst to the plant as it increases the shoot length of the plant rapidly. As it is also proven from the data that the value of the shoot length of the plant was greater in the presence (which is 5% concentration) of the urea than in absence of urea.

12.   Limitations

This investigation could not clearly explain the rate of cell division taking place within the seeds and neither could explain whether the seeds were in dormancy stage. The rate of absorption of urea might differ between the chickpea seeds and the wheat seeds. Some random errors can be caused by the ruler used to measure the distance between seed and the shoot length. While sowing the seeds in the seed bed container the depth was not considered essential which could cause a small error in the measurement of the shoot length and lead to increase in the percentage uncertainty in the shoot length of the plants.

13.   Improvements

During the experiment instead of using a ruler a device called Auxanometer should be used to measure the growth of the shoot of the plant[35]. Furthermore, a biochemical testing could be taken place to see the amount of chemicals that is resisted due to the competition and then supplying the same chemicals to the plants may give better results for the experiment.

14.   Extension

Additional investigation can be conducted by taking light as an independent variable and keeping all other variables constant and also the same chemicals or fertilisers (Urea) must be used to get the exact results. The investigation can be conducted in different intensity and wavelength of light which would prove to be beneficial for the farmers. This would provide the farmers with an estimate of the amount of light that would be appropriate for the maximum yield of the crop. Another investigation can be done to check whether the interspecific and intraspecific competition affect the seed germinating hormone gibberellic acid and amylase enzyme activity respectively.

---

[35] "Experiments on Plant Growth | Botany." Biology Discussion, Bharti d, 20 Oct. 2016, www.biologydiscussion.com/experiments/experiments-on-plant-growth-botany/57266.

15.   Bibliography

Book Source

- Minka Peeters. IBID Biology. 4th ed. Melton, Australia: pg 447 IBID Press, 2008. Print.

Web Source

- Acevedo, E., et al. Wheat Growth and Physiology. 2002,
  www.fao.org/docrep/006/y4011e/y4011e06.htm.

- "ALL INDIA COORDINATED RESEARCH PROJECT ON CHICKPEA." ALL INDIA
  COORDINATED RESEARCH PROJECT ON CHICKPEA,
  www.aicrpchickpea.res.in/.

- Beals, M., et al. "INTERSPECIFIC COMPETITION: LOTKA-VOLTERRA." INTERSPECIFIC
  COMPETITION, 12 Dec. 2017,
  www.tiem.utk.edu/~gross/bioed/bealsmodules/competition.html.

- Burchill, Shirley. "REPRODUCTION IN PLANTS." The Open Door Web Site : Biology :
  Reproduction in Plants : Conditions Needed for the Seed to Germinate,
  www.saburchill.com/chapters/chap0048.html.

- Chandrasekara, Dinara. "Germination | Seed Germination for Kids | Science Lessons for
  Kids." k8schoollessons.Com, 29 Aug. 2017, k8schoollessons.com/germination/.

- Craine, Joseph M., and Ray Dybzinski. "Mechanisms Of Plant Competition For Nutrients, Water
  And Light." *Functional Ecology*, vol 27, no. 4, 2013, pp. 833-840. *Wiley-Blackwell*,
  doi:10.1111/1365-2435.12081.
  http://onlinelibrary.wiley.com/doi/10.1111/1365-2435.12081/full

- "Experiments on Plant Growth | Botany." Biology Discussion, Bharti d, 20 Oct. 2016,
  www.biologydiscussion.com/experiments/experiments-on-plant-growth-botany/57266.

- "Factors Affecting Germination of Seeds." Factors Affecting Germination of Seeds |
  TutorVista.com,
  www.tutorvista.com/biology/factors-affecting-germination-of-seeds.

- Fan, M. et al. "Improving Crop Productivity And Resource Use Efficiency To Ensure Food Security
  And Environmental Quality In China." Journal Of Experimental Botany, vol 63, no. 1, 2011, pp. 13-
  24. Oxford University Press (OUP), doi:10.1093/jxb/err248.
  https://academic.oup.com/jxb/article/63/1/13/553113.

- *Fcnym.Unlp.Edu.Ar*, 2018,
  http://www.fcnym.unlp.edu.ar/catedras/ecopoblaciones/TP/Begon%20Capitulo%205.pdf.

- "Fertilizer Urea." Fertilizer Urea : Nitrogen : Nutrient Management : Agriculture : University of Minnesota Extension,
  www.extension.umn.edu/agriculture/nutrient-management/nitrogen/fertilizer-urea/.

- "Gilbert, Scott. "Germination". *Ncbi.nlm.nih.gov*. Web. 2 Dec. 2016.
  https://www.ncbi.nlm.nih.gov/pmc/articles/PMC5590903/

- Growth and Development Guide for Spring Wheat,
  www.extension.umn.edu/agriculture/small-grains/growth-and-development/spring-wheat/.

- Houghton, Jennie, et al. Proceedings of the Royal Society B: Biological Sciences, The Royal Society, 7 July 2013,
  www.ncbi.nlm.nih.gov/pmc/articles/PMC3673064/.

- "How to Grow Wheat | Guide to Growing Wheat." Organic & Heirloom Seeds from Heirloom Organics,
  www.heirloom-organics.com/guide/va/guidetogrowingwheat.html.

- "How to Know If Garden Seed Is Viable." Horticulture, 29 Jan. 2013,
  www.hortmag.com/weekly-tips/propagation/how-to-know-if-garden-seed-is-viable

- "India Wheat Production by Year." India Wheat Production by Year (1000 MT),
  www.indexmundi.com/Agriculture/?country=in&commodity=wheat&graph=production

- Kaya, C., Levent Tuna, A. & Alfredo, A.C.A. Acta Physiol Plant (2006) 28: 331.
  https://doi.org/10.1007/s11738-006-0029-7

- "Nitrogen." Efficient Fertilizer Use Guide Nitrogen | Mosaic Crop Nutrition,
  www.cropnutrition.com/efu-nitrogen.

- "Nitrogen." Efficient Fertilizer Use Guide Nitrogen | Mosaic Crop Nutrition,
  www.cropnutrition.com/efu-nitrogen.

- Nogueira, S.S.S., et al. Growth Analysis of Chickpea (Cicer Arietinum L.). Dec. 1994,
  www.scielo.br/scielo.php?script=sci_arttext&pid=S0103-90161994000300008.

- "Physical Properties Of Urea." *Sciencing*, 2018,
  https://sciencing.com/physical-properties-urea-6369247.html.

- "Principle of Competitive Exclusion." Encyclopaedia Britannica,
  www.britannica.com/science/principle-of-competitive-exclusion.

- "Sample Variance: Simple Definition, How to Find It in Easy Steps." Statistics How To,
  www.statisticshowto.com/sample-variance/.

- Skálová, Hana et al. "Effect Of Intra- And Interspecific Competition On The Performance Of Native And Invasive Species Of Impatiens Under Varying Levels Of Shade And Moisture." *Plos ONE*, vol 8, no. 5, 2013, p. e62842. *Public Library Of Science (Plos)*, doi:10.1371/journal.pone.0062842. http://journals.plos.org/plosone/article?id=10.1371/journal.pone.0062842

- Skálová, Hana, et al. "Effect of Intra- and Interspecific Competition on the Performance of Native and Invasive Species of Impatiens under Varying Levels of Shade and Moisture."PLoS ONE, Public Library of Science, 2013, www.ncbi.nlm.nih.gov/pmc/articles/PMC3651240/.

- Streck, Nereu Augusto et al. "Efeito Do Espaçamento De Plantio No Crescimento, Desenvolvimento E Produtividade Da Mandioca Em Ambiente Subtropical." Bragantia, vol 73, no. 4, 2014, pp. 407-415. Fapunifesp (Scielo), doi:10.1590/1678-4499.0159. http://www.scielo.br/scielo.php?pid=S0006-87052014000400009&script=sci_arttext&tlng=en

- "The Effect of Light on Germination and Seedlings." Thompson & Morgan, www.thompson-morgan.com/effect-of-light.

- Tilley, Nikki. "What Is Soil Temperature: Learn About The Ideal Soil Temperatures For Planting." Gardening Know How, 26 Nov. 2016, www.gardeningknowhow.com/garden-how-to/soil-fertilizers/determining-soil-temperature.htm.

- Tooley, Jill. "How To Clean Your Reusable Water Bottle Or Travel Mug". *Promotional Products Blog |*
Quality Logo Products (QLP). Web. 2 Dec. 2016. https://www.qualitylogoproducts.com/blog/how-to-clean-reusable-water-bottle-travelmug/.

- "Urea Formula." Math, www.softschools.com/formulas/chemistry/urea_formula/420/.

- Vogt, Deborah & Murrell, David & Stoll, Peter. (2009). Testing Spatial Theories of Plant Coexistence: No Consistent Differences in Intra-and Interspecific Interaction Distances. The American Naturalist. 175. 73-84. https://www.journals.uchicago.edu/doi/abs/10.1086/648556

- "Worksheet For How To Calculate T Test." Ncalculators.Com, 2018, https://ncalculators.com/math-worksheets/how-to-calculate-t-test.htm.

16.   Appendix

To find out the differences between the shoot lengths of the seeds in interspecific and intraspecific competition mathematical analysis was done.

Mathematical analysis includes the average and statistical differences between the shoot lengths

(i)   Average value for the shoot length measured in Table 1 was calculated

$$Average\ Value = \frac{Value\ in\ each\ trial}{number\ of\ trials}$$

An example for this can be for the wheat seeds for the distance 50 mm in the absence of urea

$$Average\ Value = \frac{168.5 + 168 + 167.5 + 168.5 + 169}{5}$$

Average Value = 168.3 mm

(ii)   Average value of the shoot length measured in Table 2 and 3 were calculated

$$Average\ Value = \frac{Value\ in\ each\ trial}{number\ of\ trials}$$

An example for this can be the chickpea seeds for the distance 10 mm in the presence of urea

$$Average\ Value = \frac{80.5 + 79.5 + 80.5 + 81 + 81.5 + 82 + 81.5 + 80.5 + 81 + 81.5}{10}$$

Average Value = 80.95 mm

(iii)   T-test was done for the Table 2 and 3 for the chickpea seeds

T-test can be done by first finding the variance by the formula

$$s^2 = \frac{\Sigma (X - \overline{X})^2}{N-1}$$ [36]

Then finding the calculated value of the t-test by using the formula

$$t = \frac{\overline{x_1} - \overline{x_2}}{\sqrt{\dfrac{S_1^{\,2}}{N_1} + \dfrac{S_2^{\,2}}{N_2}}}$$ [37]

[36] "Sample Variance: Simple Definition, How to Find It in Easy Steps." Statistics How To, www.statisticshowto.com/sample-variance/.

[37] "Worksheet For How To Calculate T Test." Ncalculators.Com, 2018, https://ncalculators.com/math-worksheets/how-to-calculate-t-test.htm.

Distances between the seeds (mm)	Concentration of urea	T test done for the seeds in the presence and absence of urea sown in different distances in the soil										C- Value	P-Value
10	presence	80.5	79.5	80.5	81	81.5	82	81.5	80.5	81	81.5	3.04E-05	2.101
	absence	79	78.5	79.5	79.5	79	79.5	80	80	79	79.5		
30	presence	90.5	91	90	91	90.5	89.5	90	90.5	89.5	90.5	0.000919253	2.101
	absence	89.5	88.5	89.5	89	87.5	89.5	88.5	88	89	88.5		
50	presence	97	96.5	97	96.5	96.5	97.5	98	97	97.5	98	8.50E-05	2.101
	absence	95.5	95.5	95	95	96	96.5	96.5	96	97	97.5		
70	presence	103.5	102.5	102.5	101	101.5	102	102.5	102	102	102.5	0.304781563	2.101
	absence	101	100.5	100.5	101.5	99.5	103.5	103	102	102.5	103		

- **Interspecific Competition:** It is the competition between the plants of different species for some limiting   resource which affects the shoot length of the plant.
- **Intraspecific Competition:** It is the competition between the plants of similar species competing for the limiting resources. It is more severe than the interspecific competition.
- **Urea:** A crystalline compound which is the main nitrogenous source. It is also used as fertilizer for the crops.
- **Sand Bed:** The fixed amount of sand in the container which is the medium on which the plants are grown.
- **Shoot length:** It is the length of the shoot developed when the seed develops into a seedling. It is the distance from the bottom to the major tip of the plant.
- **Plumule length:** It is the embryonic shoot of the plant. The shoot emerges from the plumule.
- **Percentage error:** It is the measurement of the inaccuracy in the obtained results.
- **Fertilizer:** It is a substance (chemical or natural) which is used to increase the fertility of the soil and decrease the growth time of a plant.
- **Seed Dispersal:** It is the movement of the seeds away from the plants, maybe through air or agents, to decrease the extent of competition between them.
- **Seed Dormancy:** It is a stage of seed in which the seed is prevented from germinating even under the normal atmospheric condition.
- **Auxanometer:** It is an apparatus which can measure the rate of growth of the plant.

YOUR KNOWLEDGE HAS VALUE

- We will publish your bachelor's and
  master's thesis, essays and papers

- Your own eBook and book -
  sold worldwide in all relevant shops

- Earn money with each sale

Upload your text at www.GRIN.com
and publish for free